LABORATORY NOTEBOOK

NAME

SCHOOL

ADDRESS

EMAIL

PHONE

TABLE OF CONTENT

www.ingramcontent.com/pod-product-compliance
Lightning Source LLC
Chambersburg PA
CBHW082242220526
45469CB00012B/2663
* 9 7 8 1 7 2 6 0 0 8 8 5 3 *